AF882459

Interactions Between Health and Farm-Labor Productivity

Kwadwo Asenso-Okyere
Catherine Chiang
Paul Thangata
Kwaw S. Andam

International Food Policy Research Institute
Washington, DC

January 2011

ABOUT IFPRI

The International Food Policy Research Institute (IFPRI®) was established in 1975 to identify and analyze alternative national and international strategies and policies for meeting food needs of the developing world on a sustainable basis, with particular emphasis on low-income countries and on the poorer groups in those countries. While the research effort is geared to the precise objective of contributing to the reduction of hunger and malnutrition, the factors involved are many and wide-ranging, requiring analysis of underlying processes and extending beyond a narrowly defined food sector. The Institute's research program reflects worldwide collaboration with governments and private and public institutions interested in increasing food production and improving the equity of its distribution. Research results are disseminated to policymakers, opinion formers, administrators, policy analysts, researchers, and others concerned with national and international food and agricultural policy.

Cover illustration and design: Joan Stephens, JKS Design.

DOI: http://dx.doi.org/10.2499/9780896295421

ISBN 10-digit: 0-89629-542-7
ISBN 13-digit: 978-0-89629-542-1

Contents

Table & Figures

Boxes

Acknowledgments

We gratefully acknowledge the contribution of the Bill & Melinda Gates Foundation (BMGF), which provided funds for the study that led to the publication of this report. We also thank Teunis van Rheenen, who coordinated the IFPRI service unit under which the study was conducted. We appreciate the research assistance of Daniel Ayalew Mekonnen during the latter part of the study. We also thank BMGF and IFPRI staff members who provided useful comments on earlier versions of the main report. We are also grateful to Gwendolyn Stansbury and the editorial team of the communications division of IFPRI for their assistance.

Partners and Contributors

IFPRI gratefully acknowledges the generous unrestricted funding from Australia, Canada, China, Denmark, Finland, France, Germany, India, Ireland, Italy, Japan, the Netherlands, Norway, the Philippines, South Africa, Sweden, Switzerland, the United Kingdom, the United States, and the World Bank.

Executive Summary

In the 21st century, agriculture remains fundamental to economic growth, poverty alleviation, improvement in rural livelihood, and environmental sustainability (World Bank 2007). Three-quarters of the world's poor live in rural areas, particularly in Asia and Africa (Ravallion, Chen, and Sangraula 2007), and depend on agriculture as their primary source of livelihood.

This report provides an overview of current knowledge of the impact of health issues on farm-level productivity and decisionmaking, and the impact of agriculture on health. Findings are based on a review of the relevant studies of agricultural regions throughout the developing world. Two conceptual frameworks are used to frame this research:

(1) Examining the two-way linkages between agriculture and health

(2) Tracking the pathway from a disease condition to its effects, including impacts on household decisionmaking and ultimate impacts on livelihood.

Agriculture underpins the health of rural households. It provides income that makes households resilient to health shocks; it provides food to meet their nutrient and energy needs; and it provides medicinal plants for treating ailments.

But agricultural systems can also have negative effects on health. Agricultural development may lead to environmental change with adverse health impacts: for example, irrigation dams that create suitable conditions for mosquitoes may lead to increased incidence of malaria locally. The use of agricultural inputs such as pesticides by untrained farm personnel often causes illness. Improper food harvesting and storage practices allow mycotoxins to flourish. Lack of diet diversity can lead to malnutrition. Certain animal diseases also can infect humans. Labor migration (including agricultural labor migration) can contribute to high incidence of HIV infection.

The effects of ill health on farm households include three broad impacts: absenteeism from work due to morbidity (and eventual death); family time diverted to caring for the sick; and loss of savings and assets in dealing with disease and its consequences. The long-term impacts of ill health include loss of farming knowledge, reduction of land under cultivation, planting of less labor-intensive crops, reduction of variety of crops planted, and reduction of livestock. The ultimate impact of ill health is a decline in household income and possible food insecurity—that is, a severe deterioration in household livelihood.

The research found that the household's ability to cope with a shock reflected both its asset portfolio—including human, physical, and financial assets—and its intangible social resources. Good health must be seen as both an investment and consumption asset, like agricultural production, in that it has compounding returns. Health problems, conversely, may trigger a cycle of lowered agricultural productivity and poor health. At the household level, the investment in health can improve resilience and enhance the ability to cope with emergencies, including ill health. But an investment in health in turn requires an adequate livelihood. Access to appropriate inputs (knowledge, land, tools, fertilizer, and seeds) and remunerative markets is necessary to improve the productivity, health, and resilience of farm households.

Recommendations

The results of this analysis suggest the following policy and program recommendations.

1. Include social protection and social safety net provisions in rural development policies.

Public programs to support vulnerable households should be designed to serve specific development goals, and should be well targeted and include an exit strategy (whether time-based or linked to specific objectives or results). Programs to assist rural households should pay particular attention to women, who tend to be disadvantaged in many developing countries.

2. Address ecological restoration and maintenance through policies and programs of collective action and knowledge sharing.

Encourage communities to undertake activities that restore deforestation and degraded lands using a combination of scientific and indigenous knowledge.

3. Combat health threats of widespread pesticide use and aflatoxin incidence.

Educate farmers and farm workers on the use of protective clothing against the harmful effects of pesticides. Enact legislation to enforce the safe use of pesticides and to regulate their testing, production, formulation, transportation, marketing, and disposal, in conformity with international standards. Educate rural people about aflatoxin sources and health impacts. Promote good practices in drying food commodities.

4. Collaborate more broadly through cross-sectoral, regional, and global programs.

Synergistic rural development calls for intersectoral partnerships between agriculture and health. Such partnerships will require regular monitoring and occasional impact assessments to gauge their effectiveness.

Similarly, programs to combat animal disease can benefit from regional and global cooperation in surveillance, diagnosis, and treatment. The "One Health" concept further emphasizes the interconnection of human health and animal health, as evidenced in the spread of zoonotic and pandemic diseases such as SARS and avian influenza. Global initiatives such as joint veterinary and human-health surveillance and control of zoonotic diseases should be expanded in the context of the One Health principle.

5. Invest in essential research.

Programs to support rural development would benefit from a systematic study of the interaction between health and nutrition on the one hand, and agricultural labor productivity on the other. Especially useful would be a methodology to measure farm-labor productivity at the household level, in relation to the burden of disease. More information is also needed on disease-specific impacts on farm-labor productivity—particularly the impacts of lesser-known agriculture-related diseases such as food-borne diseases, pesticide poisoning, and zoonotic diseases.

Other important topics for research include:

- The "paddies paradox": the finding that an increase in mosquito population in the vicinity of water projects does not correlate with increased malaria incidence
- The effects of climate change on human infectious diseases, and the needed mitigation and adaptation of policies and programs
- Current and potential availability of nutritious (micronutrient-rich) staple foods through biofortification

Health Threats, Productivity, and Livelihoods in Agricultural Communities

In the 21st century, agriculture remains fundamental to economic growth, poverty alleviation, improvement to rural livelihood, and environmental sustainability (World Bank 2007). Three-quarters of the world's poor live in rural areas, particularly in Asia and Africa (Ravallion, Chen, and Sangraula 2007), and depend on agriculture as their primary source of livelihood. The success of agricultural livelihoods depends on the health of the workforce, and agricultural production systems can have impacts on workers' health, nutrition, and well-being. And the labor market consequences of poor health are likely to be more serious for the poor, who are more likely to suffer from severe health problems and to be working in jobs for which strength, and therefore good health, are required (Strauss and Thomas 1998). This report provides an overview of current knowledge of the impact of health issues on farm-level productivity and decisionmaking, and the impact of agriculture on health, based on a review of the relevant studies of agricultural regions throughout the developing world.

Agriculture underpins the health of rural households. It provides income that makes households resilient to health shocks; it provides food to meet their nutrient and energy needs; and it provides medicinal plants for treating ailments. But agricultural systems can also have negative effects on health. Agricultural development often leads to environmental change that can have adverse health impacts: for example, irrigation dams or water storage receptacles that create suitable conditions for the breeding of mosquitoes may lead to increased incidence of malaria in the neighborhood. The use of agricultural inputs such as pesticides by untrained farm personnel often causes illness. Improper food harvesting and storage practices allow mycotoxins toxic to humans and animals alike to flourish. Lack of diet diversity and/or overconsumption of certain food items (oils, fats, and the like) can lead to malnutrition. Certain animal diseases also can infect humans, with grave consequences. Labor migration (including agricultural labor migration) can contribute to high incidence of HIV infection.

Two frameworks are used here to study the interaction between health and farm-labor productivity: (1) a framework for understanding the two-way linkages between agriculture and health; and (2) a framework for tracking the pathway from a disease condition to its effects, its impacts on household decisionmaking, and its ultimate impacts on human livelihood.

Agriculture and disease affect one another in a bidirectional manner: agricultural development projects may affect disease causation; and diseases that afflict farmers may negatively affect the farmers' productivity (or require adjustments in labor allocation). Figure 1 presents a framework, developed by Hawkes and Ruel (2006a), for understanding the linkages between agriculture and health. The entire agricultural supply chain—agricultural producers, agricultural systems, and agricultural outputs—has implications for health, mediated by critical intermediary processes: the labor process, environmental change, income generation, and access to food, water, land, and health-related services.

Figure 1—Framework for linkages between agriculture and health

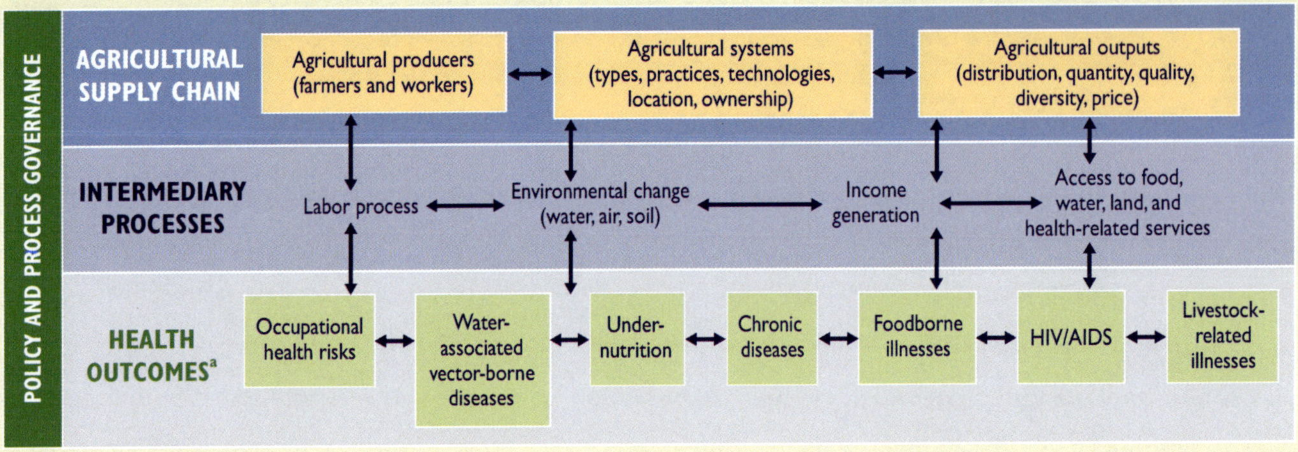

Source: Hawkes and Ruel (2006a).

Figure 2—Conceptual framework for the impact of illness/disease on agriculture

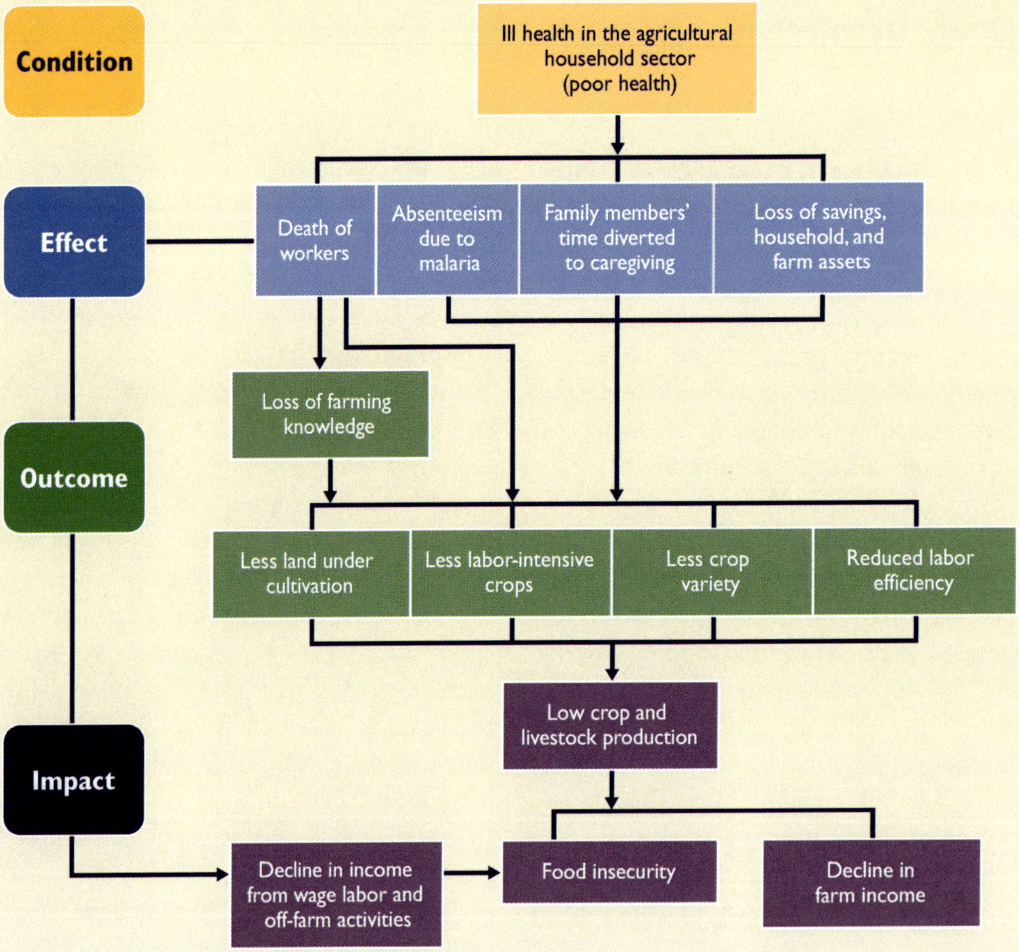

Source: Adapted from Negin (2005) and Asenso-Okyere et al. (2009).

Agricultural development and practice can exacerbate the incidence of disease, through an interaction with disease vectors and parasites.

We look first at the impact of ill health on household productivity and well-being, and then turn to the health implications of agricultural products and practices. Figure 2 presents a conceptual framework for analyzing the impacts of ill health on agriculture.

Impacts of Illness on Household Livelihood

Poor health (from whatever cause) can inflict great hardships on households, including debilitation, substantial monetary expenditures, loss of labor, and sometimes death. More broadly, the health and nutritional status of adults affects their ability to work, and thus underpins the welfare of the household, including the children's development.

Treatable conditions often go untreated because of lack of access to healthcare. Many rural areas do not have clinics; the sick must be carried on the backs of young men or on bicycles to the nearest clinic. Moreover, clinics in rural areas often lack adequate equipment or trained health personnel, and in many countries they require payment before providing service. In the absence of health insurance, rural people are often unable to afford healthcare of any kind.

Poor health in turn affects agricultural production. Illness impairs the farmer's ability to innovate, experiment, and implement changes, and to acquire technical information available through extension activities. Healthcare expenses may consume resources that otherwise might be used to purchase improved seed, fertilizer, equipment, or other inputs. Households with sick members are less able to adopt labor-intensive techniques. As a counterpoint, health threats also affect the demand for agricultural output. Malnutrition and disease patterns influence market demand for food, in terms of quantity, quality, diversity, and the price people are able or willing to pay.

The long-term household impacts of ill health include loss of farming knowledge, reduction of land under cultivation, planting of less labor-intensive crops, reduction of variety of crops planted, and reduction of livestock. The ultimate impact of ill health is a decline in household income and possible food insecurity—that is, a severe deterioration in household livelihood.

Low labor productivity is a distinguishing characteristic of developing-country agriculture. As shown in Figure 3, labor productivity (measured in terms of agriculture value-added per worker) is quite low in low-income or developing countries, as compared to high- and middle-income countries, which rely more on farm machinery than labor. Rampant poor health among the adult population in developing countries contributes to low productivity.

Figure 3—Agriculture value-added per worker by income group, 2000 and 2005

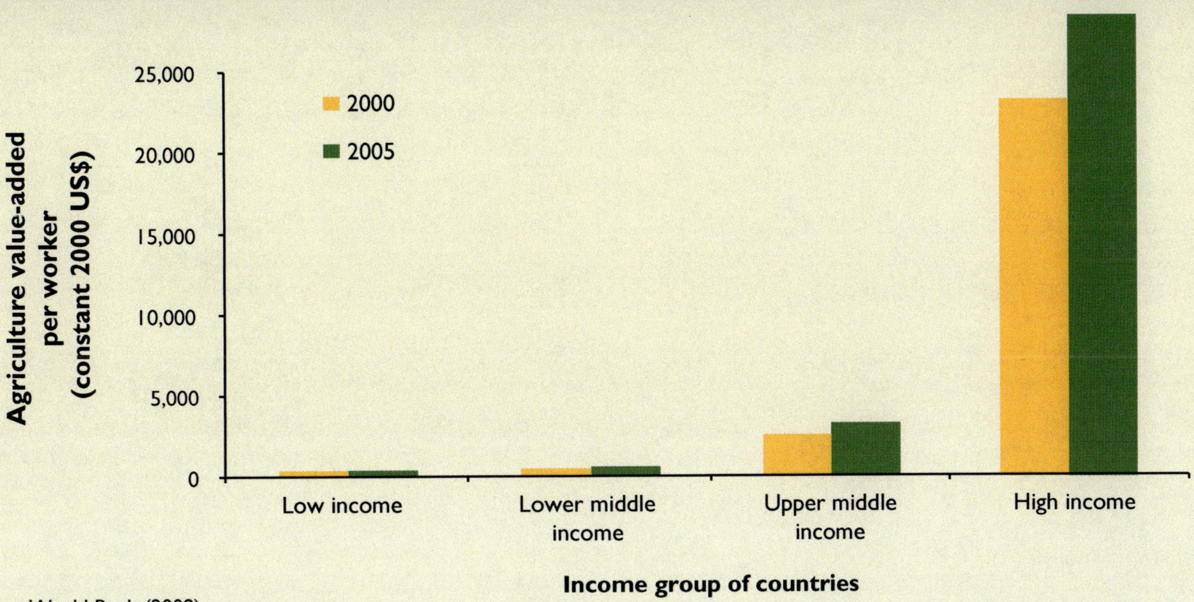

Source: World Bank (2008).

Absenteeism, Burden of Caregiving, and Loss of Assets

This report discusses the effects of ill health on farm households in terms of three parameters:

- absenteeism from work due to morbidity (and eventual death)
- family time diverted to caring for the sick
- loss of savings and assets in dealing with disease and its impacts

The literature documents substantial absenteeism due to illness. In Oyo State in Nigeria, the estimated average number of workdays lost per year due to malaria was 64 days in agrarian households. In Leyte province in the Philippines, 45.4 days in a year were lost to schistosomiasis. In India, the average number of days lost to tuberculosis (TB) was 83 days per year.

One study analyzed the effects of illness on agricultural households in 22 districts in rural Kenya, a country with one of the highest rates of HIV infection (Yamano and Jayne 2004). HIV-related deaths led to losses in cultivated area, crop production levels, household assets, and off-farm income (see Figure 4). Following the death of the household head (male or female), households tend to lose some members to migration or off-farm activities. The death of the household head leads to a reduction in the amount of crop land devoted to high-value crops, as well as reduced crop output. This is especially so in the case of the death of male heads of households, who tend to have the knowledge of high-value crop production. Thus, households turn to low-value crops when they lose a male household head, and they also lose some farm labor due to the departure of other household members. As might be expected, household assets tend to decline following the death of household members, possibly due to the sale of assets or through inheritance.

Caregiving responsibilities also take time away from productive work. A study in the Free State province of South Africa found that AIDS and other HIV-related diseases require an average of about 2,700 caregiving hours a year from household members (equivalent to about 113 person-days). In Ethiopia, 42 person-days per year were devoted to taking care of a child with malaria. Often children are withdrawn from school for varying periods to help with caregiving or household chores (especially girls), and some take up employment to help support their families. (Note that depriving children of their education in order to work is a contravention of the International Labour Organisation (ILO) Fundamental Conventions 138 and 182.)

Figure 4—Impacts of illness on agricultural productivity in rural Kenya

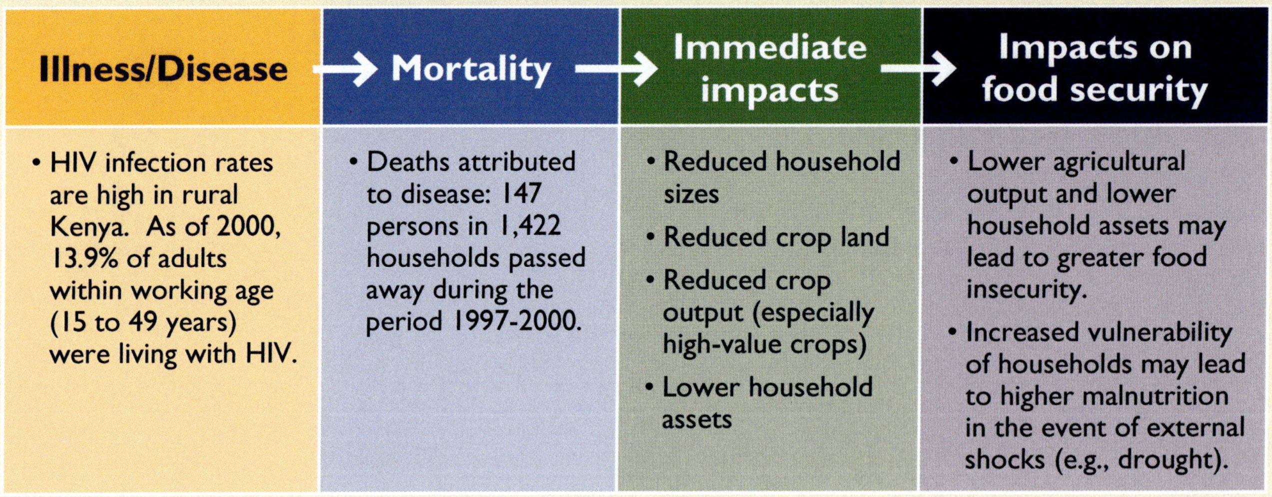

Source: Yamano and Jayne (2004).

Box 1—Economic Burden of Illness on Households

A multicountry study of onchocercal skin disease in Ethiopia, Sudan, and Nigeria found that people with this disease spent 15 percent of their annual income on health-related expenditures. A review of studies on the economic burden of illness for households in developing countries, focusing on malaria, TB, and HIV, found that these illnesses imposed regressive cost burdens on poorer patients and their families; TB, HIV, and AIDS cost households more than 10 percent of income. In Côte d'Ivoire, healthcare costs specific to the person with AIDS accounted for almost 80 percent of the household healthcare budget. In the Rungwe district of Tanzania, rising medical expenses significantly increased the probability of a household's falling below the poverty line. High medical expenses further undermine the household ability to purchase, maintain, and replace essential farm inputs. A study in Uganda found that it was difficult for HIV-affected households to adopt recommended agronomic practices that boost the production and quality of produce because they had limited funds to invest in farm inputs and implements.

Sources: WHO (1997), Russell (2004), Tibaijuka (1997), Booysen and Bachmann (2002), Pitayanon et al. (1997), Bechu (1998), Mwakalobo (2003), NAADS (2003).

The combination of healthcare costs and loss of productivity create a significant burden on households affected by severe or chronic illness (Box 1).

Death of a household member of course implies permanent loss of labor. In Thailand, 35 percent of households with a member who died from AIDS suffered a decline of 48 percent in household income. In Zimbabwe, death of a household head due to AIDS caused an average 61 percent reduction in maize output. Lost labor may be replaced by bringing in extended family members, who may be unemployed or underemployed; by withdrawing children from school to assist on the farm; or by hiring labor if the household can afford to do so. Little research has been conducted on the effect on the generational transfer of agricultural knowledge from the death of one (or both) parents.

Given the labor-intensive nature of agricultural systems in developing countries, disease and the associated loss of labor can have significant consequences. Farm households attempt to address the shortage of labor through various methods, such as reducing the area under cultivation or narrowing the range of varieties planted on the farm. Beyond the direct impacts due to loss of labor, illness undermines long-term agricultural productivity in a number of ways. When illness leads to long-term incapacitation, households may resort to withdrawing savings, selling important assets (such as jewelry, textiles, breeding animals, farm equipment, and land), withdrawing children from school, or reducing the nutritional value of their food consumption. All of these emergency responses can have adverse effects on the long-term labor productivity of household members. Such household adjustments have been shown to affect female members disproportionately.

Major Health Threats in Agricultural Regions

The diseases that have had the most severe impact on livelihoods in the developing world are HIV/AIDS, malaria, and tuberculosis. In addition, there are chronic effects from diseases more closely related to agriculture: soil- and water-borne diseases, mycotoxins, and zoonotic diseases. Cardiovascular diseases also represent a growing health threat. Finally, the impact of malnutrition may have effects on productivity as well as on health.

HIV/AIDS

Sub-Saharan Africa is the most heavily HIV-affected region worldwide, representing 67 percent of infections. Women and girls continue to be disproportionately affected by HIV in the region, with

Table 1—Life expectancy with and without AIDS in selected African countries, 2000 and 2010

Country	2000			2010		
	With	Without	Years Lost	With	Without	Years Lost
Botswana	39.3	70.5	31.2	29.0	73.2	44.2
Ethiopia	45.2	56.1	10.9	42.1	60.0	18.0
Kenya	48.0	64.9	16.9	44.3	68.4	24.1
Nigeria	53.6	57.8	4.2	38.9	64.0	26.0
South Africa	51.1	65.7	14.6	35.5	68.3	32.8
Swaziland	40.4	57.7	17.3	29.7	61.5	31.8
Zimbabwe	37.8	69.9	32.1	32.5	72.8	40.3
Zambia	37.2	58.7	21.5	38.9	72.8	33.9

Source: U.S. Bureau of the Census (2010).

Box 2—HIV and AIDS and Absenteeism

Dick Huck Farm (in Mt. Darwin, Zimbabwe) was once a prosperous sweet potato farm employing more than 500 workers. But a few years ago, labor problems set in. "We started witnessing most of our specialized workers developing full blown AIDS and skipping work, and our cash flows dropped drastically," said farm manager Josphat Jonga. In South Africa, voluntary testing of 1,500 employees on 18 commercial farms recently revealed that 28.5 percent are infected with HIV. One quarter of Swaziland's workforce is absent because of HIV and AIDS, according to the International Monetary Fund. Elsewhere in Africa, farmers are having to source migrant workers as their own staff fall sick and many are reporting a serious loss of agricultural skills.

Source: CTA (2009).

women accounting for approximately 60 percent of all estimated HIV infections (UNAIDS 2008; Garcia-Calleja, Gouws, and Ghys 2006). Of the roughly 2 million AIDS-related deaths in 2008, Sub-Saharan Africa accounted for 72 percent. HIV infection in the region has resulted in a substantial decrease in life expectancy (Table 1), even after several years of investments to control HIV and AIDS.

Globally, the number of cases of HIV infection has grown by 20 percent in the past decade, with estimates as high as 35 million in 2008; 50 percent of the infected people are women (UNAIDS and WHO 2009). Households and farms affected by AIDS suffer major losses of productivity (see Box 2). AIDS threatens household livelihoods, community well-being, and national food security and poverty reduction efforts.

Malaria

Malaria is one of the most prevalent and challenging infectious diseases affecting developing countries. Intensive malaria is confined to the tropical and subtropical zones, and poverty may promote its transmission. It is endemic in 91 countries whose population makes up 40 percent of the world's population, and it is responsible for more than 1 million deaths per year, mostly children (McCarthy et al. 2000). While malaria is nonfatal in the majority of cases, it results in recurrent debilitating bouts of illness and limits individuals' ability to work productively (Cole and Neumayer 2006). Because malaria mostly affects children, a significant amount of adults' (women's) time is diverted to caring for sick children. Repeated bouts of malaria in agrarian households cause a decline in farm output and farm income, resulting in food insecurity and an increase in poverty (ESPD 2005).

The threat of malaria is increasing globally, as a result of increasing population movements into malarious regions, changing agricultural practices (including the building of dams and irrigation systems), deforestation, the weakening of public health systems in poor countries, and increasing resistance of the parasite to drug treatment. Long-term climate changes may also play a role, including more pronounced El Niño cycles and global warming (Sachs and Melaney 2002). Fortunately, new forms of treatment and control are being explored, including two especially promising approaches: artemisinin-based combination therapies (ACTs), as well as synthetic substitutes still under development; and a biological control method that uses larvae-eating fish to control mosquito populations in rainfed pools.

Tuberculosis

TB was declared a global health emergency in 1993, and today it threatens one-third of the world's population (WHO 2009b). Africa has the highest incidence rate (363 cases per 100,000 people) and the highest mortality rate; the Asian region had the largest number of cases (63 percent of cases worldwide, in 2007). There have been improvements in disease control, however: global TB prevalence and death rates per 100,000 people declined between 1990 and 2006—from 296 to 206 (prevalence), and from 28 to 25 (mortality; WHO 2009c).

Soil- and Water-Borne Diseases

Helminth infections (roundworm, whipworm, hookworm), caused by soil-transmitted helminths (STHs) and schistosomes, are among the most prevalent afflictions in areas of poverty in the developing world. Transmission is associated with poor sanitation and lack of clean water, allowing contamination of environments with egg-carrying feces. While rarely fatal, STH infections have chronic and insidious effects on health and nutritional status (Stephenson et al. 2000; Stolzfus et al. 1997). The greatest prevalence is in the Americas, China and East Asia, and Sub-Saharan Africa. These last three regions (excluding the Americas) report a total of 1.2 billion cases of roundworm, 795 million cases of whipworm, and 750 million cases of hookworm (de Silva et al. 2003). Approximately 300 million people globally suffer from severe morbidity as a result of these diseases (Crompton 1999; Montresor et al. 2002).

Schistosomiasis (also known as Bilharziasis) is a water-borne parasitic disease caused by trematode flatworms of the genus Schistosoma (WHO 2009d). It is second to malaria as the most devastating parasitic disease in tropical countries. An estimated 200 million people in 74 countries are infected with the disease (Carter Center 2009)—half of them in Africa. While schistosomiasis has a relatively low mortality rate, it has a high morbidity rate, causing severe debilitating illness that can damage internal organs and, in children, can impair growth and cognitive development, and thus their productivity and earning ability as adults. Schistosomiasis is often associated with water resource development projects, such as dams and irrigation schemes, as well as rivers.

Mycotoxins

Mycotoxins are toxic secondary metabolites of fungal origin that can contaminate agricultural commodities before or after harvest. When ingested, inhaled, or absorbed through the skin, mycotoxins cause lowered performance, sickness, or death of humans and animals (Wagacha and Muthomi 2008). The five agriculturally important toxins from fungi are aflatoxins, fumonisins, ochratoxin A, zearalenone, and deoxynivanelol (Bankole and Adebanjo 2003). Aflatoxins in particular pose serious health, economic, and agricultural problems in developing countries (Bankole and Adebanjo 2003). Aflatoxin chemical poisons, produced mainly by the fungus Aspergillus flavus, may contaminate such foods as cassava, peanuts, corn, rice, cottonseed, and other grains. Contamination is associated with high stress factors during plant growth, late harvesting of crops, high ambient humidity (preventing thorough drying), and poor storage conditions.

Sub-Saharan Africa and Southeast Asia show high levels of aflatoxin exposure (Jolly et al. 2007). Research in West Africa found detectable amounts of aflatoxin in the blood samples of more than 98 percent of children and adults (Montesano et al. 1997; Wild and Turner 2002). Aflatoxin contamination has been linked with the high incidence of liver cancer in Africa (Oettle 1964; Bababunni et al. 1978). Children in Togo and Benin who ate foods contaminated with aflatoxins exhibited stunted growth and underweight, symptoms associated with malnutrition (Gong et al. 2002). Aflatoxin exposure in West Africa at the time of weaning has been associated with impaired child growth and decreased immune response (Turner et al. 2003). An estimated 40 percent of disease-associated loss of productivity in developing countries is due to diseases exacerbated by aflatoxins (Miller 1996).

The good news is that collaborative research by the African Agricultural Technology Foundation, Kenya and the United States Department of Agriculture, and the International Institute for Tropical Agriculture has identified a natural fungus, found in Nigeria, that can significantly reduce concentration of aflatoxins in maize (*Rural* 21 2009). Another promising strategy uses food additives to protect farm animals from aflatoxin, and thus reduce human exposure (Williams et al. 2004).

Zoonotic Diseases

Zoonotic diseases are defined as human diseases that are acquired from (or transmitted to) any other vertebrate (Box 3). More than three-quarters of the human diseases that are currently new, emerging, or re-emerging are caused by pathogens originating either from animals or from products of animal origin (FAO/WHO/OIE 2004). The emergence of zoonotics is largely the result of human activities: increased migration; increased intensification of market-oriented smallholder livestock systems; and adaptive agropastoral and pastoral systems serving large local and regional markets. Intensification of animal production tends to increase disease risk.

Zoonotic diseases may also be contracted from nonagricultural activities. In some parts of west and central Africa's farming areas, bush meat is the main source of meat for many households (Nasi 2007). The handling of freshly butchered bush meat, in particular primates, brings about a risk of transmission of new zoonoses. Pathogens that do not cause disease in their natural hosts can do so in their new hosts, or may evolve to do so. The Simian Immunodeficiency Virus

Box 3—Zoonotic Diseases

The World Health Organization defines zoonoses as diseases caused by infectious agents that are naturally transmitted between animals and humans. Some well-known zoonoses are salmonellosis, swineherd's disease (caused by Leptospira spp.), brucellosis, hepatitis E, HIV, bovine spongiform encephalopathy (BSE)—and its zoonotic form, the variant Creutzfeldt-Jakob disease (vCJD)—Rift Valley fever (RVF), anthrax, adult meningitis (caused by Streptococcus suis), and influenza.

The currently emerging zoonoses include SARS, West Nile virus, and highly pathogenic avian influenza. Outbreaks of these diseases are related to several factors: increasing population, globalization, trade in exotic pets, and the close intermingling of animals and humans in urban settings. In Africa, movements of domestic and wild animal populations are important in the spread of the diseases, which pose an especially serious threat to the fragile economies of Sub-Saharan Africa, with its nascent livestock, dairy, and poultry industries. A very serious emerging zoonotic disease in South Asia is Nipah virus disease, transmitted by Pteropus bats, which causes high levels of fatality in humans and in pigs.

Source: National Research Council (2008).

(SIV) and Human Immunodeficiency Virus (HIV) are thought to have separately crossed over from African monkeys and apes to humans; Ebola hemorrhagic fever is lethal to gorillas and chimpanzees as well as to humans (POST 2005).

The consequences of livestock diseases include direct economic costs, such as the loss of animal production and products, as well as indirect costs, such as the loss of trade, markets, and jobs (OIE 1999; Le Gall 2006). In the 1980s, an outbreak of foot-and-mouth disease caused a 30 percent loss of milk production in the Kenyan dairy farming sector (Le Gall 2006). In 1997–98, Rift Valley fever virus in East Africa undermined birth of calves and milk production; milk exports declined by 75 percent (Le Gall 2006). The 2003–04 outbreak of highly pathogenic avian influenza in Southeast Asia resulted in more than 140 million dead or destroyed birds and economic losses exceeding US$10 billion (OIE/FAO/WHO 2007). Since late 2003, the H5N1 strain of avian influenza has been responsible for 4,544 documented outbreaks in poultry farms in 36 countries. These outbreaks were associated with 269 human cases and 163 fatalities as of January 2007 (World Bank 2007). While avian influenza has generally been dealt with by culling birds, health authorities are now trying to identify and address possible sources of infection farther up the supply chain, according to David Nabarro, the United Nations senior coordinator for avian and pandemic flu (*Daily Monitor* 2010).

Cardiovascular Diseases

Cardiovascular diseases (CVD) are newly becoming an important health hazard in developing countries, especially in urban areas. While they are by no means specific to agricultural settings, they can have severe consequences for productivity and livelihoods. By 2001, CVD had become the leading cause of death in the developing world outside Sub-Saharan Africa (where HIV-related disease has higher mortality; Mathers et al. 2006; WHO 2002). Contributing factors include changing diets—from coarse grains, fruits, and vegetables to finer grains, meat, oil, fat, and sugar—along with increasingly sedentary lifestyles, sometimes accompanied by smoking and heavy consumption of alcohol. The health consequences, including heart disease, high cholesterol, hypertension, and diabetes, often require daily medication. Treatment costs can force even rather well-to-do households into poverty, in the absence of health insurance. Conservative estimates in Brazil, China, India, Mexico, and South Africa indicate that each year at least 21 million years of future productive life are lost because of CVD (Mathers et al. 2006).

Malnutrition

The burden of high medical costs at the household level often results in reduced consumption, including of basic needs such as food (Pitayanon et al. 1997). A survey in Côte d'Ivoire found that per capita food consumption in AIDS-afflicted households was half that of other households (Bechu 1998). A study of 300 households in two districts in Zimbabwe distinguished between coping-level households[1] and acute-level households;[2] the large majority were acutely affected, at 277 households. Only 23 percent of the acutely affected households had three meals per day, with less diversity in their diets than coping-level households (Food Security Network of Zimbabwe 2007). Substantially reduced consumption can lead to malnutrition, a weakened immune system, and particularly an increase in AIDS patients' susceptibility to opportunistic diseases. In general, the World Health Organization identifies malnutrition as "the single most important risk factor for disease." The Integrated Agriculture-Aquaculture project, discussed in Box 8, represents an important initiative for increasing household nutrition, which is particularly advantageous to communities affected by HIV.

Assessing the Productivity Effects of Ill Health and Malnutrition

High labor productivity can play a significant role in economic development, and it is the only sustainable (non-inflationary) path to higher wages (Ukoha 2000). Improvement in agriculture labor productivity is critical in developing countries, where agriculture is a major source of employment and livelihood.

Various studies have been undertaken in Sierra Leone, India, Sri Lanka, the Philippines, Ethiopia, and Mali to assess the impact of health and nutrition on productivity of agricultural workers. The general conclusion is that poor health (defined broadly in terms of nutritional and health status) has significant impacts on farm productivity. The studies are not directly comparable, however, as they use different definitions for nutritional status, nor is it clear how health status was measured.

Many studies have measured farm-labor productivity as output per unit of time per farm worker. It can also be measured as value of goods and services produced in a period of time, divided by the hours of labor used to produce the goods and services. Note that labor productivity may not completely capture worker efficiency: a farm can boost output per worker by introducing machinery or adopting a new technology. Similarly, a farm can lose output per worker if a disease strikes the workforce. A more comprehensive measure of an economy's (or farm's) use of resources is "total factor productivity" (TFP), an index of the efficiency of use of both capital (land) and labor. TFP is calculated as the percentage increase in output that is not accounted for by changes in the volume of inputs (capital, including land, and labor). Thus, TFP decreases if disease lowers the efficiency of labor, holding other factors constant. Although TFP is a more comprehensive measure of resource allocation, the analysis of TFP is best done at the macroeconomic level, not at the household level. This review therefore focuses on *farm-labor productivity*, a term widely used in the empirical literature. However, the full impact of diseases on productivity is only partially captured by measuring days of work missed.[3] Future research would benefit from a methodology that can correctly capture farm-labor productivity at the household level, accurately reflecting the burden of disease and malnutrition.

Health Impacts of Agricultural Production

Agricultural production is a determinant of health, primarily through the consumption of food produced and through intermediary processes related to income and labor. In addition to providing some or all of the household's food needs, agriculture provides income for farmers and farm laborers, enabling access to food, water, land, information and education, and health-related services—in short, enabling nutrition and good health.

But agricultural labor can also affect the household's nutritional status adversely, both through high expenditure of energy and by usurping time that might be spent on childcare, food preparation, and other nutrition-related activities—or, if inefficiently utilized, detracting from more rewarding income-generating activities.

Finally, agricultural labor exposes producers to a range of occupational health hazards, such as accidents, diseases, and poisoning from pesticides.

Nutrition and Other Crop Values

Agricultural output also affects health through availability of quality food. Quality and diversity of food influence access to micronutrients and dietary diversity that enhance resistance to disease. Certain types of food, such as fats and oils, and foods which contain dangerous substances like aflatoxins, have given rise to diet-related chronic diseases.

Agriculture also contributes medicinal plants that help treat diseases. For instance, because of the high treatment costs and difficulties with access, only a small percentage of households with people living with HIV or AIDS are currently using pharmaceuticals and supplements. Instead, they depend on local capacities and resources, including plant-based medicine sourced from the forest (Willumsen and Kettaneh 2005; FAO 2003). Examples of medicinal plants include the bark of *Prunus africana* tree, used in the treatment of prostate disorders; *Artemisia annua* (sweet wormwood), used in treating malaria; and the African tree *Melaleuca alternifolia* (tea tree), which contains an antifungal substance that combats *Candida albicans*, the bacteria responsible for fungal skin problems and mycosis (a condition that commonly affects the eyes of AIDS patients). Medicinal plants constitute a fundamental component of traditional healthcare systems in rural communities throughout Africa. WHO estimates that about two-thirds of the world's population, and 80 percent of Africa's population, sometimes use herbal or traditional medicine.

Pesticide Use

As pesticide use has increased in developing countries, so has pesticide poisoning in farmers (Box 4). Farmers use stronger concentrations of pesticides, with increased frequency of application, and they mix several pesticides together to combat pesticide resistance by pests (Chandrasekera et al. 1985; WRI 1998). Due to lack of training, many farmers contract pesticide-

Box 4—Pesticide Use and Farm Worker Health

While pesticides can increase agricultural productivity, there is growing concern about the health impacts of their use. When handled improperly, agricultural chemicals are toxic to humans and other species. Deaths due to unintentional poisoning from exposure to pesticides are estimated at 355,000 people yearly, two-thirds of whom are in developing countries.

Research results of a potato farming community in Ecuador revealed a rate of 171 pesticide poisonings per 100,000 people during 1991–92, 10 times the level reported by the Ministry of Health. Days of recuperation averaged 11 days of lost labor wages. Median lost-time indirect cost was US$8.33 per case—more than five days' income (the local agricultural wage was about US$1.50 per day). Hospital records showed numerous cases of poisoning of women and children, in addition to the main pesticide applicators. Pesticide residues were found on workers applying the pesticides, on children who had been in the field, and on various surfaces inside the farm household.

Negative health and economic impacts of pesticides can be minimized through training and information campaigns on pesticide use. In Nicaragua, farmers trained in appropriate pesticide use suffered lower exposure after two years and had higher net returns than those who had not been trained.

Sources: Cole et al. (2000), Yanggen et al. (2004), World Bank (2007).

related diseases (Antle and Pingali 1994). Farm workers may not use protective clothing or equipment, whether because they are not aware of the dangers, the clothes are unavailable or unaffordable, or there are no regulations to enforce using them. Many of the negative health impacts resulting from pesticide use can be mitigated if protective measures are taken and recommended methods are followed when mixing and applying the chemicals.

Consequences of large-scale pesticide use include hormone disruption and immune suppression (Straube et al. 1999), as well as damage to skin and eyes, and even death. Prolonged exposure to pesticides can cause chronic health problems: cardiopulmonary problems, neurological and hematological symptoms, and adverse dermal effects (Davies, Freed, and Whittemore n.d.; Spear 1991). The Food and Agriculture Organization estimates that approximately 3 million people are poisoned and 200,000 die from pesticide use each year (FAO 2000). A more recent estimate by the World Bank puts deaths caused by pesticide poisoning at 355,000 annually (see Box 4).

Improper use of pesticides also has less obvious impacts on productivity. In Tanzania, a study of vegetable farmers reported that 68 percent of farmers who used pesticides reported feeling sick after routine pesticide application (Ngowi et al. 2007). In Zimbabwe, pesticide acute symptoms significantly increased the direct cost of illness in cotton growers. Moreover, producers spent two to four days of the growing season recuperating from illnesses attributed to pesticides (Maumbe and Swinton 2003). A study of two rice-producing regions of the Philippines, and one of potato farms in Ecuador, found that pesticide use has a negative effect on farmer health, while conversely good farmer health has a significant positive effect on productivity (Antle and Pingali 1994; Antle et al. 1998). The Ecuador study concluded that a reduction in the use of the principal insecticide (carbofuran) could raise productivity as well as improve farmers' health.

It is therefore important to consider health effects in the economic analysis of pesticide adoption in agricultural production. One study finds that in rice production, the value of crops lost to pests is invariably lower than the cost of pesticide-related illness and the associated loss in farmer productivity (Rola and Pingali 1993). Failure to consider health effects may result in underestimating the economic benefit of pest-resistant crops or organic farming (using only biological control of pests).

Pesticides also contaminate drinking water and food crops, especially fruits and vegetables receiving higher doses of pesticides, thus posing serious health hazards to consumers (Pimental et al. 1992). According to the U.S. Food and Drug Administration, approximately 35 percent of the foods purchased by consumers have detectable levels of pesticide residues, and 1–3 percent of the foods have pesticide residue levels above the legal tolerance levels.

Labor Migration

Migrant farmers are essential to cocoa production in West Africa. Farmers in Ghana and Côte d'Ivoire have relocated to acquire land for cocoa farming. During the long dry season in the Sahel region of West Africa, people migrate to the cocoa-growing areas to find temporary jobs as farm laborers. Migrants also include women who provide sexual services to the migrant farm workers. Any diseases, especially sexually transmitted diseases, then spread to migrants' households in the area of origin. Poor health and malnutrition affect the work performance of agricultural producers, reducing productivity and income (Hawkes and Ruel 2006a; Hawkes and Ruel 2006b).

Child Labor

Inadequate labor availability to meet production needs is one aspect underlying the use of child labor on family cocoa farms in West Africa, a practice that affects children's health and education (Barrientos et al. 2008). (Child labor is distinguished from child work, in which a child accompanies a parent to undertake light farm work that does not affect his or her schooling; Asuming-Brempong et al. 2007).

Changing Land and Water Use Patterns

Agricultural systems also affect human health through the environmental changes produced in water, soil, and air. Deforestation resulting from road construction, logging, and farming has opened up opportunities for increased consumption of bush meat, with the associated threat of zoonoses (Patz et al. 2004).

Certain features—such as crop rotation, the presence of livestock, and the proximity of villages to fields and water sources—create conditions for contracting water-borne vector diseases (World Bank 2007). Similarly, agricultural development has increased the need for crop irrigation, which creates conditions that favor the breeding of parasitic vectors and facilitates disease transmission. In the southern Nile Delta, for example, an increase in soil moisture associated with irrigation development, following the construction of the Aswan Dam, caused a rapid rise in the mosquito population and an increase in the number of cases of Bancroftian filariasis (Harb el al. 1993; Thompson et al. 1996). In Ethiopia, a study reported that the introduction of microdams for irrigation resulted in a seven-fold increase in the incidence of malaria (Ghebreyesus et al. 1999). Onchocerciasis and trypanosomiasis, which cause river blindness and sleeping sickness respectively, are vector-borne parasitic diseases that can emerge in response to changing land-use and water-management practices. Research in Malaysia reported that expansion and changes in agricultural practices were intimately associated with the emergence of Nipah virus in the country (Chua et al. 1999). Also, ponds used to collect water for irrigation in peri-urban agriculture breed mosquitoes, increasing the incidence of malaria for nearby households (Box 5).

Project planning needs to recognize and address the potential consequences of changes in land and water use, designing appropriate intervention activities. Discussions between agriculture and health policymakers and professionals can help to identify and assess the externalities of these projects and minimize their negative impacts.

Food-Borne Pathogens

Agriculture's major output, food, can carry diseases caused by contamination by pathogens during agricultural production. Consumption of milk contaminated by *Mycobacterium bovis*, which is present in animals in most developing countries, has long been regarded as the principal mode of TB transmission from animals to humans (Acha and Szyfres 1987). *M. bovis* and *M. tuberculosis* have been found in milk samples in Ethiopia, Nigeria, and Egypt, highlighting the serious public health implications of potentially contaminated milk and milk products in developing countries (WHO 1994; Idrisu and Schnurrenberger 1977; Nafeh et al. 1992). One way of reducing the effects of the contaminants is to boil the milk before consumption; however, in many milk-producing communities, milk is consumed fresh.

Coping Strategies

Research has found that a household's ability to cope with a shock reflects both its asset portfolio—including human, physical, and financial assets—and its intangible social resources. Good health should be seen as both an investment and consumption asset, like agricultural production, in that it has compounding returns. Health problems, conversely, may trigger a cycle of lowered agricultural productivity and poverty. (See Box 6 for effects of different types of health problems.)

Box 5—Peri-urban Agriculture and Malaria

To feed the increasing urban population, urban and peri-urban agriculture has intensified, especially in Africa. Farmers cultivate undeveloped land around the cities to produce and supply vegetables to the city dwellers. Cultivation is done all year round, and availability of water is vital. Water supply becomes a constraining factor, so most farmers construct wells or harvest water and store it in dugouts and bunds. These water storage receptacles provide favorable aquatic habitats for mosquitoes. People who live around these gardens easily contract malaria throughout the year, if they are not properly protected by bednets, pesticides, or other control measures. A study in Kumasi, Ghana, to assess the impact of irrigated urban agriculture on malaria transmission revealed higher adult anopheline mosquito densities in peri-urban and urban agricultural locations, with more reported malaria episodes than in the nonagricultural locations in the city. The study found high levels of parasitemia among children living in communities closer to agricultural sites. It is therefore important for policymakers in agriculture and health to be cognizant of this relationship when devising agricultural development policies and strategies. More research is needed to provide detailed information on the health impacts of urban agriculture.

Source: Afrane (2003).

Box 6—Four Levels of Health Threat

Acute mild or moderate illness. Studies in Sri Lanka, Vietnam, and Bangladesh found that minor illnesses are a significant shock to poor and vulnerable households, especially those with few assets, forcing them into debt or to deplete their few remaining assets to meet minor healthcare expenses.

Recurring illness. Although the disease burden of recurring illness is larger than that of common illness, few studies have looked at its link to impoverishment at the household level.

Chronic and long-term illness. Research in Thailand found that the financial impact of TB on poor households was devastating, with 15 percent of poor households selling property and 10 percent taking out loans to finance treatment costs. Studies conducted in South Asia also found that a large percentage of TB-affected households become indebted due to medical expenses.

Terminal and steadily deteriorating health. HIV infection imposes an enormous cost burden on many households in developing countries. Research in Zimbabwe reports that more than 60 percent of households have borrowed money to cover the direct costs of the disease; about one-third have reduced expenditure on basic needs; and between 20 and 30 percent have sold assets. Studies from various parts of Sub-Saharan Africa indicate that HIV and AIDS, through loss of income and loss of productive asset sales, cause a process of household impoverishment.

Sources: Russell (2004); Southern Africa Partnership Programme (2005).

An investment in personal health takes the form of preventing disease exposure, consuming adequate food and good nutrition, and seeking appropriate healthcare. At the household level, such investments improve resilience and enhance the ability to cope with emergencies, including ill health. The investment in health in turn requires an adequate livelihood. Access to appropriate inputs (knowledge, land, tools, fertilizer, and seeds) and remunerative markets is essential to improving the productivity, health, and resilience of farm households.

Coping Strategies for Four States of Health

Households affected with *common illnesses*, particularly those with young children, often need to finance medical costs by cutting back spending. They may use savings, pawn jewelry, or borrow the money needed. Studies in Sri Lanka, Vietnam, and Bangladesh found that even minor illnesses are a significant shock to poor and vulnerable households, especially those with few assets. They may go into debt or deplete their few remaining assets to meet minor healthcare expenses.

Recurring illnesses, such as malaria, impose chronic hardships on households. They use their cash reserves and savings, or they may have to borrow, sell assets and labor, and rely on social networks to manage the costs of illness. The disease burden of recurring illness is larger than that of common illness, but few studies have looked at its impact on household impoverishment.

In developing countries where social safety nets are limited or non-existent, *chronic conditions* such as diabetes, hypertension, and tuberculosis impose high costs over time, if regular treatment is required and if the sick are recurrently incapacitated. Research in Thailand found that the financial impact of TB for poor households was devastating: 15 percent of poor households sell property, and 10 percent take out loans to finance treatment costs. Studies conducted in South Asia also found that a large percentage of TB-affected households become indebted due to medical expenses. Overall, the cost burden is high, forcing unsustainable strategies that reduce assets, increase debt, and create vulnerability to future shocks.

HIV infection, a condition which causes steadily *deteriorating health*, imposes an enormous cost burden on many households in developing countries. Research in Zimbabwe reports that more than 60 percent of households have borrowed money to cover the direct costs of the disease; about one-third have reduced their expenditures on basic needs; and between 20 and 30 percent have sold assets. Studies from various parts of Sub-Saharan Africa detail the loss of income and loss of productive assets caused by HIV and AIDS, and the resulting household impoverishment. A study by the Southern Africa Partnership Program discusses the example of livestock sales: "Sales of chickens, goats, or cattle are classic coping strategies that households all over Sub-Saharan Africa employ. Some level of livestock sales is normal and does not result in increased poverty. At a certain point, however, household livestock holdings reduce to the level where they are no longer sustainable. At this point, livestock sales as a coping strategy become erosive." Substantial sales of livestock also have implications for crop production, due to reduced availability of draught power and manure, with implications for a household's future production. (Russell 2004; Southern Africa Partnership Programme 2005.)

Enhancing Household Productivity

Rural households can be assisted with well-designed strategies to make the most of their available labor resources. Two such strategies are home gardens and conservation farming techniques.

Home gardens, with their associated crop diversity, provide an excellent foundation to enhance household food security and nutrition. Home gardening concentrates on smaller-sized family (or community) gardens, including traditional, neglected, and underutilized crops. Fruit trees are an important component. Improving home gardening requires the optimal use of local agrobiodiversity; with careful species mixes, such gardens can produce food year-round. Crops are selected for their food and nutritional value rather than their market value. Home gardens tend to appeal most strongly to women, who are often in charge of selecting, cooking, and growing the family's food needs.

Conservation farming techniques make optimal use of available labor. For example, land preparation tasks are shifted to the dry season, when there are few other agricultural tasks. Other strategies focus on increasing the stock and the health of animal assets within the small-scale farm sector, as well as the stock of equipment such as plows and harrows. Enhancing farmers' incentives and ability to acquire draft animals and equipment can also help alleviate the crucial labor burden of land preparation (Jayne et al. 2005).

In Siavonga, Zambia, the "basin planting" method (a water-harvesting technology) has been introduced to smallholder farmers who lack access to draft animal power for land preparation. The basins are made using a hand hoe during the dry season, when labor is available, requiring about the same level of effort as plowing before planting. At the same time, cover crop production is also promoted as a way of maintaining soil moisture, reducing runoff, increasing infiltration, reducing erosion, and increasing or maintaining organic matter throughout the year.

In Tanzania, the use of crop residues and cover crops (*Dolichos lablab, Crotalaria, Mucuna,* and *Canavalia*), as well as dead weeds and mulch, is a documented method to suppress weeds and reduce labor demand for weeding (Shetto and Owenya 2007). Additionally, leguminous cover crops, such as *Mucuna*, fix nitrogen, hence improving soil fertility without additional labor input. By increasing water retention capacity of soils, this technology also helps to improve yields in dry years. However, basin planting and cover crops have yet to be widely adopted in Zambia and Tanzania.

The Importance of Fish

Fish is in many areas the only accessible and affordable source of animal protein for poor households. It offers fats (macronutrients) and micronutrients such as iron, iodine, zinc, calcium, and vitamins A and B. In Malawi, 70 percent of protein comes from fish. However, per capita fish consumption has decreased over the past decades, from 14 kg per person per year in the 1970s to about 4 kg today (Nagoli et al. 2009).

Agricultural investments have yet to adequately respond to the needs, and mobilize the strengths, of vulnerable households, such as those affected by debilitating illness (Nagoli et al. 2009). HIV-affected households are among the worst afflicted by food and nutrition insecurity and reduced income levels. Improved agricultural technologies tend to impose increased burdens on households, in terms of labor, capital investment, time, institutional support, and planning capacity, which are often lacking in households stressed by illness. Moreover, HIV-affected households spend substantial time, labor, and other resources to care for their sick members who no longer participate in economic activities (Gillespie 2006).

Fish, and in particular cultured fish (in small ponds), can play a mitigating role in the combined crisis of HIV and food insecurity in Malawi (Bene and Heck 2005), especially through technological innovations that take into account the requirements of the affected households.

Box 7—Conservation Farming: Hand Jab Planter

In Brazil and Paraguay, the Hand Jab Planter (HJP) is widely used as a hand tool for planting into soil cover. Once the technique is mastered, the HJP reduces labor energy demand, as it requires only one person for planting instead of three traditionally (for digging the hole, planting, and closing the hole). The cost is estimated at approximately US$10 to manufacture in Arusha, Tanzania. However, its uptake in Africa has been minimal.

Sources: Jayne et al. (2005); Shetto and Owenya (2007).

Box 8—Integrated Agriculture-Aquaculture (IAA) Project for HIV-Affected Households

The WorldFish Center, in collaboration with World Vision Malawi, carried out a one-year World Bank–funded project in Chingale-Zomba district, covering 37 villages (in Traditional Authority Mlumbe), in 2005–2006. The aim of Integrated Agriculture-Aquaculture (IAA) project was to identify constraints on HIV-affected households' adoption of fish farming, and to adapt technologies and practices to boost their fish production and utilization. The IAA project introduced 134,000 fingerlings of *Tilapia rendalli* into new and rehabilitated ponds. *Tilapia rendalli* feeds on rice or maize bran as supplementary feed and on large amounts of vegetation. This fish species grows to market size (100 grams) in 4–6 months (Nagoli et al. 2009).

The yield from the ponds averaged 1500 kg/hectare, totaling 21 metric tons of fresh fish. The local per capita supply of fresh fish was estimated to have increased by 150percent—from about one kilogram of fresh fish per person. About 1,200 households benefited, of which 60 percent were women-headed households affected by HIV. The investment in farm inputs was minimal, as farmers were encouraged to use farm wastes and crop by-products to feed their fish. The provision of additional water from the ponds in the dry season also expanded crop and vegetable production. This allowed farmers to grow additional valuable crops like bananas and guava on the perimeter of their ponds.

Developing customized technologies to serve HIV-affected households is a critical component in an all-inclusive strategy for economic growth. IAA can dramatically improve the ability of families to cope with the effects of HIV and AIDS. The technology ensures nutritional benefits for the sick and improves household income and food security.

Source: Nagoli et al. (2009).

Conclusion

This report shows that a household's ability to cope with a shock (such as illness) is greatly influenced by its asset portfolio, including human, physical, and financial assets as well as intangible social resources. Health must be viewed as both an investment and a consumption asset, much like agriculture. When both health and agriculture thrive, a reinforcing cycle of health can result. When either suffers, the cycle becomes a negative spiral of lowered agricultural productivity and diminished health.

While agriculture provides essential food and fiber, agricultural development and practice can exacerbate the incidence of disease through interaction with disease vectors and parasites. When disease (from any source) afflicts farmers, their productivity is reduced and households are trapped in poverty, with an unacceptable standard of living. The impacts of ill health and consequent household adjustments seem to affect females disproportionally.

Households and individuals can invest in health through good lifestyles, by avoiding exposure and taking preventive measures, including adequate food and good nutrition as well as appropriate healthcare. Such steps will improve households' resilience and enhance their ability to cope with emergencies such as ill health. Good health is an asset for agriculture, by improving productivity. Conversely, agriculture is an asset that contributes to good health and resilience. Farmers with access to appropriate inputs (including knowledge, land, tools, fertilizer, and seeds) as well as remunerative markets will increase their productivity and earn good incomes that allow them to thrive nutritionally, acquire more assets (including health), and become more resilient.

Notes

1. Coping-level household was defined as a household in a vulnerable situation but still able to cope; a household is considered coping level if its Household Vulnerability Index (HVI) was less than 0.33 (Food Security Network of Zimbabwe 2007).

2. Acute-level household was defined as a household that has been hit so hard it badly needs assistance to the degree of acute healthcare. A household is considered acute level if HVI is greater than or equal to 0.33 and less than or equal to 0.66 (Food Security Network of Zimbabwe 2007).

3. Similarly, wage levels do not accurately capture changes in productivity, as they may not reflect labor market competition for various reasons, including scarcity of labor.

References

Acha, P.N., and B. Szyfres. 1987. Zoonotic tuberculosis. In *Zoonosis and communicable diseases common to man and animals*. 2nd edition. Scientific Publication No. 503. Washington, DC: Pan American Health Organization/World Health Organization.

Afrane, Y.A. 2003. Does irrigated urban agriculture influence the transmission of malaria in the city of Kumasi, Ghana? *Acta Tropica* 89: 125–134.

Antle, J.M., D.C. Cole, and C.C. Crissman. 1998. Further evidence on pesticides, productivity and farmer health: Potato production in Ecuador. *Agricultural Economics* 18: 199–207.

Antle, J.M., and P.L. Pingali. 1994. Pesticides, productivity, and farmer health: A Philippine case study. *American Journal of Agricultural Economics* 76 (3): 418–30.

Asenso-Okyere, K., F.A. Asante, J. Tarekegn, and K.S. Andam. 2009. *The linkage between agriculture and health.* International Food Policy Research Institute Discussion Paper 00861. Washington, DC: International Food Policy Research Institute.

Asuming-Brempong, S., D.B. Sarpong, P. Amoo, and K. Asenso-Okyere. 2007. *Pilot labour survey in cocoa production in Ghana*, Ministry of Manpower, Youth and Employment, Accra: Government of Ghana.

Bababunni, E.A., A.O. Uwaifo, and O.Bassir. 1978. Hepatocarcinogens in Nigeria foodstuffs. *World Review of Nutrition and Dietetics* 28: 188–209.

Bankole, S.A., A Adebanjo. 2003 Mycotoxins in west Africa: Current situation and possibilities for controlling it. *African Journal of Biotechnology* 2 (9):254–263.

Barrientos, S., K. Asenso-Okyere, S. Asuming-Brempong, D. B. Sarpong, N.A. Anyidoho, R. Kaplinsky and J. Leavy. 2008. *Mapping sustainable production in Ghanaian cocoa.* London: Cadbury.

Bechu, N. 1998. The impact of AIDS on the economy of families in Côte d'Ivoire: Changes in consumption among AIDS-affected households. In *Confronting AIDS: Evidence from the developing world*, ed. M. Ainsworth, L. Fransen, and M. Over. Brussels: European Commission.

Bene, C. and S. Heck. 2005. Fish and food security in Africa. *NAGA* 28(3&4):8–13. Malaysia: WorldFish Center.

Booysen, F., and M. Bachmann. 2002. HIV/AIDS, poverty, and growth: Evidence from a household impact study conducted in the Free State Province, South Africa. Presented at the Annual Conference of the Center for Study of African Economies. St. Catherine's College, March 18–19, Oxford.

Carter Center. 2009. Schistosomiasis control program. http://www.cartercenter.org/health/schistosomiasis/index.html.> Accessed on December 11, 2009.

Chandrasekera A.I., A. Wettsinghe, and S. L. Amarasiri. 1985. Pesticide usage by vegetable farmers. Paper presented at Annual Research Conference ISTI, Gannoruwa, Sri Lanka.

Chua, K.B., K.J. Goh, K.T. Wong, A. Kamarulzaman, P.S. Tan, T.G. Ksiazek, et al. 1999. Fatal encephalitis due to Nipah virus among pig farmers in Malaysia. *Lancet* 354: 1257–1259.

Cole, M., and E. Neumayer. 2006. The impact of poor health on total factor productivity. *The Journal of Development Studies* 42 (6): 918–938.

Crompton, D.W. 1999. How much helminthiasis is there in the world? *Journal of Parasitology* 85: 397–403.

CTA (Technical Centre for Agricultural and Rural Cooperation ACP-EU). 2009. Agriculture and Health: Fighting back. Spore 142 (August 2009). http://spore.cta.int/index.php?option=com_content&task=view&lang=en&id=934&catid=9. Accessed on December 11, 2009.

Daily Monitor. 2010. Vol. XVI, No. 090. pp. 3. Addis Ababa, Ethiopia, April 21.

Davies, J.E., V.H. Freed, and F.W. Whittemore. n.d. An agrochemical approach to pesticide management: Some health and environmental considerations. Miami: University of Miami.

de Silva, N. R., S. Brooker, P. J. Hotez, A. Montresor, D. Engles, and L. Savioli. 2003. Soil-transmitted helminth infections: Updating the global picture. *Trends in Parasitology* 19: 547–51.

ESPD. 2005. *Enhancing health systems: Malaria's negative impact in Africa*. Economic and Social Policy Division. Addis Ababa: Economic Commission for Africa.

FAO (Food and Agricultural Organization). 2000. Project concept proposal. *HEAL: Health in ecological agriculture learning*. Prepared by the FAO Programme for Community IPM in Asia. Rome: FAO.

———. 2003. HIV/AIDS and agriculture: Case studies from Namibia, Uganda and Zambia. Rome: FAO.

FAO, WHO (World Health Organization), and OIE (World Organisation for Animal Health). 2004. Report of the WHO/FAO/OIE joint consultation on emerging zoonotic diseases, May 3–5, Geneva, Switzerland.

Food Security Network of Zimbabwe. 2007. *Impact, of HIV and AIDS on rural agricultural production and food security in Chivi and Makoni districts*. Harare, Zimbabwe: Food Security Network of Zimbabwe.

Garcia-Calleja, J.M., E. Gouws, and P.D. Ghys. 2006. National population based HIV prevalence surveys in Sub-Saharan Africa: Results and implications for HIV and AIDS estimates. *Sexually Transmitted Infections* 82 (Supp. 3): iii64–iii70.

Ghebreyesus, T.A., M. Haile, K.H. Witten, A. Getachew, A.M. Yohannes, M. Yohannes, et al. 1999. Incidence of malaria among children living near dams in northern Ethiopia: community-based incidence survey. *British Medical Journal* 319: 663–666.

Gillespie, S. 2006. *Understanding the links between agriculture and health. Agriculture and HIV/AIDS*. Focus 13, Brief 7, May. Washington DC: International Food Policy Research Institute.

Gong, Y.Y., K.K. Cardwell, A. Hounsa, S. Eggal, P.C. Turner, A.J. Hall, and C.P. Wild. 2002. Dietary aflatoxin exposure and impaired growth in young children from Benin and Togo: A cross-sectional study. *British Medical Journal* 325: 20–21.

Harb, M., R. Faris, A.M. Gad, O.N. Halez, R. Ramzy, and A.A. Buck. 1993. The resurgence of lymphatic filariasis in the Nile delta. *Bulletin of the World Health Organization* 71: 49–54.

Hawkes, C., and M. T. Ruel. 2006a. The links between agriculture and health: An intersectoral opportunity to improve the health and livelihoods of the poor. *Bulletin of the World Health Organization* 84 (12): 985–991.

———. 2006b. Overview. 2020 Vision Focus Brief No. 13. In *Understanding the links between agriculture and health*, eds. C. Hawkes and M.T. Ruel. Washington, DC: International Food Policy Research Institute.

Idrisu, A., and P. Schnurrenberger. 1977. Public health significance of bovine tuberculosis in four northern states of Nigeria: A mycobacteriologic study. *Nigerian Medical Journal* 7: 384–7.

Jayne, T. S., M. Villareal, P. Pingali, and G. Hemrich. 2005. HIV/AIDS and the agricultural sector: Implications for policy in eastern and southern Africa. *Journal of Agricultural and Development Economics* 2 (2): 158–181.

Jolly P.E., Y. Jiang, W.O. Ellis, R.T. Awuah, J. Appawu, O. Nnedu, J.K. Stiles, J.S. Wang, O. Adjei, C.M. Jolly, and J.H. Williams. 2007. Association between aflatoxin exposure and health characteristics, liver function, hepatitis and malaria infections in Ghanaians. *Journal of Nutrition and Environmental Medicine* 16 (3-4): 242–257.

Le Gall, F. 2006. Economic and social consequences of animal diseases. World Bank. <http://go.worldbank.org/JRQAM0N2P0>. Accessed December 2009.

Mathers, C.D., A.D. Lopez, and C.J.L Murray. 2006. The burden of disease and mortality by condition: Data, methods, and results for 2001. In *Global burden of disease and risk factors*, ed. A.D. Lopez, C.D. Mathers, M. Ezzati, D.T. Jamison, and C.J.L Murray. New York: Oxford University Press.

Maumbe, B.M., and S.M. Swinton. 2003. Hidden health costs of pesticides use in Zimbabwe's smallholder cotton growers. *Social Science and Medicine* 57:559–1571.

McCarthy, F.D., H. Wolf, and Y. Wu. 2000. *The growth costs of malaria*. Working Paper No. 7541. New York: National Bureau of Economic Research (NBER).

Miller, J.D. 1996. Mycotoxins. In *Proceedings of the workshop on mycotoxins in food in Africa, November 6-10, 1995 at Cotonou, Benin*, ed. K.F. Cardwell. Benin: International Institute for Tropical Agriculture.

Montesano, R., P. Hainuat, and C.P. Wild. 1997. Hepatocellular carcinoma: From gene to public health. *Journal of the National Cancer Institute*. 89: 1844–1851.

Montresor, A., D. W. T. Crompton, T. W. Gyorkos, and L. Savioli. 2002. *Helminth control in school-age children: A guide for managers of control programmes*. Geneva: World Health Organization.

Mwakalobo, A. 2003. *Implications of HIV/AIDS on rural livelihoods in Tanzania: The case of Rungwe district*. Unpublished manuscript.

NAADS (National Agricultural Advisory Services). 2003. The impact of HIV/AIDS on the agricultural sector and rural livelihoods in Uganda. Baseline Report.

Nafeh, M.A., A. Medhat, A-G Abdul-Hamed, Y.A. Ahmad, N.M. Rashwan, and G.T. Strickland. 1992. Tuberculosis peritonitis in Egypt: The values of laparoscopy in diagnosis. *American Journal of Tropical Medicine and Hygiene* 47: 470–477.

Nagoli. J., E.M. Phiri, E. Kambewa, and D. Jamu. 2009. *Adapting integrated Agriculture-Aquaculture for HIV and AIDS-affected households: The case of Malawi*. The WorldFish Center Working Paper 1957. Penang, Malaysia: The WorldFish Center.

Nasi, R. 2007. *The price of a wild trade*. Spore 130. Wageningen, The Netherlands: Technical Centre for Agricultural and Rural Development.

National Research Council. 2008. *A study of emerging technologies in agriculture to benefit farmers in Sub-Saharan Africa and South Asia*. Washington, DC: National Research Council.

Negin, J. 2005. Assessing the impact of HIV/AIDS on economic growth and rural agriculture in Africa. *Journal of International Affairs* 58(2): 267–81.

Ngowi, A.V.F., T.J. Mbise, A.S.M. Ijani, L. London, and O.C. Ajayi. 2007. Smallholder vegetable farmers in Northern Tanzania: Pesticides use, practices, perceptions, cost and health effect. *Crop Protection* 26: 1617–1624.

Oettle, A.G. 1964. Cancer in Africa, especially regions south of the Sahara. *Journal of the National Cancer Institute* 33: 383–439.

OIE (World Organisation for Animal Health). 1999. *International animal health code*. OIE ed. Paris, 468 pp.

OIE, FAO, and WHO. 2007. *The global strategy for prevention and control of H5N1 Highly Pathogenic Avian Influenza (HPAI)*. March.

Patz, J.A., P. Daszak, G.M. Tabor, A.A. Aguirre, M. Pearl, J. Epstein, N.D. Wolfe, A.M. Kilpatrick. J. Foufopoulos, D. Molyneux, and D.J. Bradley. 2004. Unhealthy landscapes: Policy recommendations on land use change and infectious disease emergence. *Environmental Health Perspective* 112 (10): 1092–1098.

Pimental D, H. Acquay, M. Biltonen, P. Rice, M. Silva, J. Nelson, V. Lipner, S. Giordano, A. Horowitz, and M. D'Amore. 1992. Environmental and human costs of pesticide use. *Bioscience* 42: 750–760.

Pitayanon, S., S. Kongsin, and W. Janjaroen. 1997. The economic impact of HIV/AIDS mortality on households in Thailand. In *The economics of HIV and AIDS: The case of South and South East Asia*, ed. D. Bloom and P. Goodwin. Delhi: Oxford University Press.

POST. 2005. The bushmeat trade. The UK-Parliamentary Office of Science and Technology, February, No. 236. London.

Ravallion, M., S. Chen, and P. Sangraula. 2007. New evidence on the urbanization of global poverty. Policy Research Working Paper No. 4199. Washington, DC: World Bank. <http://econ.worldbank.org/docsearch>.

Rola, A.C., and P.L. Pingali. 1993. Pesticides, rice productivity, and farmers' health—An economic assessment. Los Baños, Laguna, Philippines: World Resources Institute and International Rice Research Institute.

Rural 21. 2009. Maize to help prevent night blindness. *Rural 21: The International Journal for Rural Development* 43 (4).

Russell, S. 2004. The economic burden of illness for household in developing countries: A review of studies focusing on malaria, tuberculosis, and human immunodeficiency virus/acquired immunodeficiency syndrome. *American Journal of Tropical Medicine and Hygiene* 71 (Suppl 2): 147–155.

Sachs, J., and P. Melaney. 2002. The economic and social burden of malaria. *Nature* 415.

Shetto, R., and M. Owenya, ed. 2007. *Conservation agriculture as practised in Tanzania: Three case studies. Nairobi. African Conservation Tillage Network, Centre de Coopération Internationale de Recherche Agronomique pour le Développement*. Rome: Food and Agriculture Organization of the United Nations.

Southern Africa Partnership Programme. 2005. *Food security and HIV and AIDS in Southern Africa*. Johannesburg: ActionAid International.

Spear, R. 1991. *Recognized and possible exposure to pesticides. Handbook of pesticide toxicology. Vol. 1, General principles*. New York: Academic Press.

Stephenson, L. S., M. C. Latham, and E. A. Ottesen. 2000. Malnutrition and parasitic helminth infections. *Parasitology* 121 (Suppl.): S23–28.

Stoltzfus, R. J., M. L. Dreyfuss, H. M. Chwaya, and M. Albonico. 1997. Hookworm control as a strategy to prevent iron deficiency anemia. *Nutrition Reviews* 55: 223–232.

Straube E., W. Straube, E. Kuger, M. Bradatsch., M. Jacob-Meisel, and H.J. Rose. 1999. Disruption of male sex hormones with regard to pesticides: Pathophysiological and regulatory aspects. *Toxicol Lett* 107:225–231.

Strauss, J., and D. Thomas. 1998. Health, nutrition, and economic development. *Journal of Economic Literature* XXXVI: 766–817.

Thompson D.F., J.B. Malone, M. Herb, R. Fairs, O.K. Huh, A.A Buck, et al. 1996. Bancroftian filariasis distribution and diurnal temperature differences in the southern Nile delta. *Emerging Infectious Diseases Journal* 2:234–235.

Tibaijuka, A. K. 1997. AIDS and economic welfare in peasant agriculture: Case studies from Kagabiro village, Kagera region, Tanzania. *World Development* 25 (6): 963–975.

Turner, P.C., S.E. Moore, A.J. Hall, A.M. Prentice, and C.P. Wild. 2003. Modification of immune function through exposure to dietary aflatoxin in Gambian children. *Environmental Health Perspectives* 111: 217–220.

Ukoha, O.O. 2000. Determinant of labor productivity on small-holder sole crop farms: A case of waterleaf enterprise (*talinum triangulare*). *Nigeria Journal of Agribusiness and Rural Development* 1:3.

UNAIDS (United Nations Programme on HIV/AIDS). 2003. HIV/AIDS, Gender and Food Security. HIV/AIDS and Gender: Fact Sheet. Geneva, UNAIDS.

———.2008. *Report on the global AIDS epidemic.* Geneva: UNAIDS.

UNAIDS and WHO (World Health Organization). 2009. AIDS epidemic update December 2009. Geneva: UNAIDS.

Wagacha, J.M., and J.W. Muthomi. 2008. Mycotoxin problem in Africa: Current status, implications to food safety and health and possible management strategies. *International Journal of Food Microbiology* 124: 1–12.

WHO (World Health Organization). 1994. *Report of WHO meeting on zoonotic tuberculosis (Mycobacterium bovis), with the participation of FAO, Mainz, Germany*. Geneva: World Health Organization. Unpub. Document WHO/CDS/VPH/94.137

———. 1997. *Economic Impact of Onchocercal Skin Disease (OSD): Report of a Multi-country Study.* TDR Applied Field Research Report. Geneva: WHO.

———. 2002. *Integrated management of cardiovascular risk.* Geneva: World Health Organization, CVD Program.

———. 2009b. Tuberculosis Fact Sheet. <http://www.who.int/mediacentre/factsheets/fs104/en/.> Accessed on December 11, 2009.

———. 2009c. *World Health Statistics 2009.*Geneva: WHO.

———.2009d. <http://www.who.int/topics/schistosomiasis/en/>. Accessed on December 11, 2009.

Wild, C.P., and P.C. Turner. 2002. The toxicology of aflatoxin as a basis for public health decisions. *Mutagenesis* 17: 471–481.

Williams, J.H., T.D. Phillips, P.E. Jolly, J.K. Stiles, C.M. Jolly, and D. Aggarwal. 2004. Human aflatoxicosis in developing countries: a review of toxicology, exposure, potential health consequences, and interventions. *American Journal of Clinical Nutrition* 80:1106–1122.

Willumsen J., and A. Kettaneh. 2005. Nutrition, HIV and AIDS. HIV and AIDS key issues guide. Health and Development Information Team. DFID Health Resources Centre, London.

World Bank. 2007. *World Development Report 2008: Agriculture for Development.* Washington, DC: World Bank.

———. 2008. World Development Indicators. Washington, DC: World Bank.

World Resources Institute. 1998. *World resources, 1998/1999.* Oxford, UK: University of Oxford Press.

Yamano, T., and T. S. Jayne. 2004. Measuring the impact of working-age adult mortality on small-scale farm households in Kenya. *World Development* 32 (1): 91–119.

About the Authors

Kwadwo Asenso-Okyere is the director of the International Food Policy Research Institute's (IFPRI) Eastern and Southern Africa Regional Office, Addis Ababa, Ethiopia.

Catherine Chiang is a research analyst in IFPRI's Environment and Production Technology Division, Washington, DC.

Paul Thangata is a research fellow in IFPRI's Eastern and Southern Africa Regional Office, Addis Ababa, Ethiopia.

Kwaw S. Andam is a young professional at the World Bank, Washington, DC.

Made in the USA
Las Vegas, NV
03 May 2026

46252451R00021